A FIRST LOOK AT FROGS, TOADS AND SALAMANDERS

By Millicent E. Selsam and Joyce Hunt
ILLUSTRATED BY HARRIETT SPRINGER

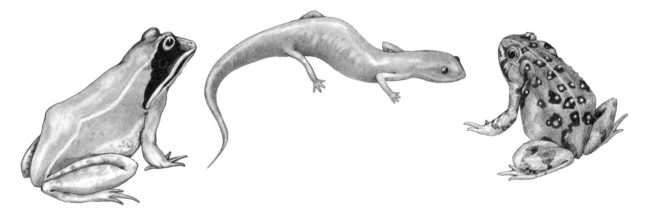

WALKER AND COMPANY ❋ NEW YORK

For Jim, who gave us the hippo

First published in the United States of America
in 1976 by the Walker Publishing Company, Inc.

Published simultaneously in Canada by Fitzhenry &
Whiteside, Limited, Toronto.

Trade ISBN: 0-8027-6243-3

Reinf. ISBN: 0-8027-6244-1

Library of Congress Catalog Card Number: 75-36019

Printed in the United States of America.

10   9   8   7   6   5   4   3   2

A *FIRST LOOK AT* SERIES

Each of the nature books for this series is planned to develop the child's powers of observation and give him or her a rudimentary grasp of scientific classification.

Frogs, toads and salamanders are amphibians.
What is an amphibian?

An amphibian is an animal that lives both on land
and in water.

This is an animal that lives on land and in water.
Is it an amphibian?

No, it is an otter.
Otters have hair.
Amphibians do not.

This is another animal that lives on land
and in water.
Is it an amphibian?

No, it is a duck.
Ducks have feathers.
Amphibians do not.

This is another animal that lives on land
and in water.
Is it an amphibian?

No, it is an alligator.
Alligators have scales.
Amphibians do not.

This is still another animal that lives on land
and in water.
Is it an amphibian?

Yes, it is an amphibian because—
It doesn't have hair like an otter.
It doesn't have feathers like a duck.
It doesn't have scales like an alligator.
Most amphibians have smooth, moist skins.

Also, an amphibian lays its eggs in water
and spends the early part of its life there.

A young amphibian is called a *tadpole.*
At first it has no legs
but it has a long, fish-like tail.
It breathes through gills.
The tadpole changes as it grows.
Hind legs appear first. Then front legs appear.
In frogs and toads the tail gets smaller and smaller
until it disappears.

EGGS

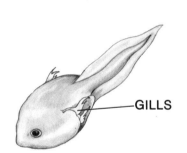

GILLS

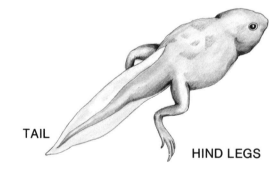

TAIL

HIND LEGS

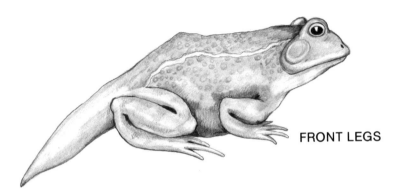

FRONT LEGS

Usually the gills disappear, too.
Meanwhile, lungs develop.
The adult amphibian is now able to breathe
the air on land.

Salamanders are amphibians with tails.

Frogs and toads are amphibians without tails.

You can tell frogs apart by the pattern
on their skins.

SPRING PEEPER

GRAY TREE FROG

PICKEREL FROG

Find the frog with round spots.
Find the frog with square spots.
Find the frog with a mask across its eyes.
Find the frog with an "X" on its back.
Find the frog with a star on its back.

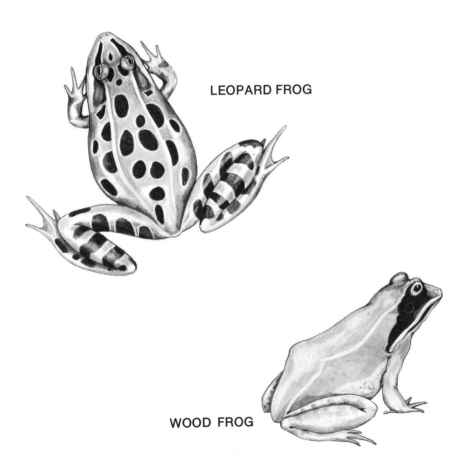

LEOPARD FROG

WOOD FROG

Some frogs have stripes.

The Swamp Tree frog has dark stripes down its back.

The Green Tree frog has a light stripe down each side and along its legs.

The Sheep frog has a light stripe down the middle of the back.

Which is which?

Sometimes size is a clue.

The Bullfrog is big. It can be 8 inches long.
The Green frog is smaller. It gets to be
only 3½ inches long.

GREEN FROG

BULLFROG

A puzzle:
Here is a small, young Bullfrog
and a large adult Green frog.
They are the same size and look very much alike.
How do we tell the difference now?

The Green frog has two folds of skin down its back.
The Bullfrog does not.

There are other things to look for, too.

Look for the frog with pupils like long slits.
Cats have this kind of eye, too.
(This frog also has spurs on its hind feet
which it uses in digging.)

Look for the frog with a pointed head.
(It also has a fold of skin
across the back of its head.)

Look for the frog with sticky pads on its feet.

Look for the frog that has a fold of skin
across the back of its head and down its sides.

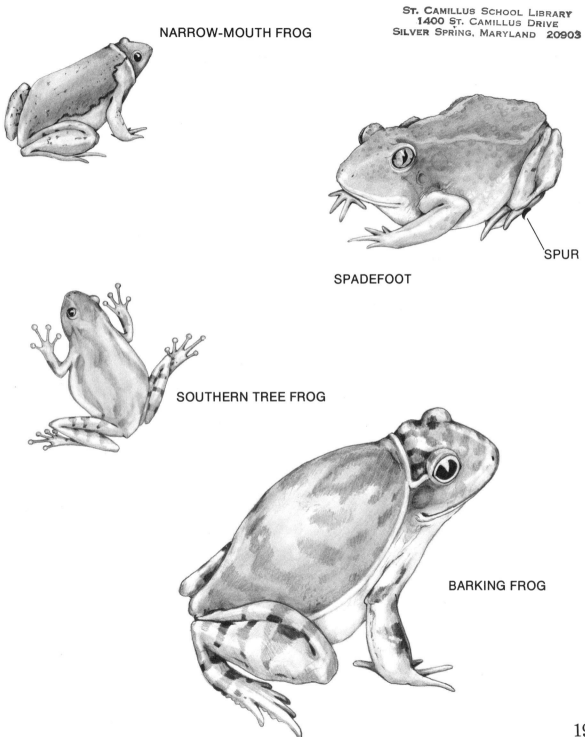

NARROW-MOUTH FROG

SPUR

SPADEFOOT

SOUTHERN TREE FROG

BARKING FROG

Here is a toad.   Here is a frog.

How can we tell the difference?

Toads have dry, warty skins   Frogs have smooth, slimy skins
and short, fat bodies.   and longer, thinner bodies.

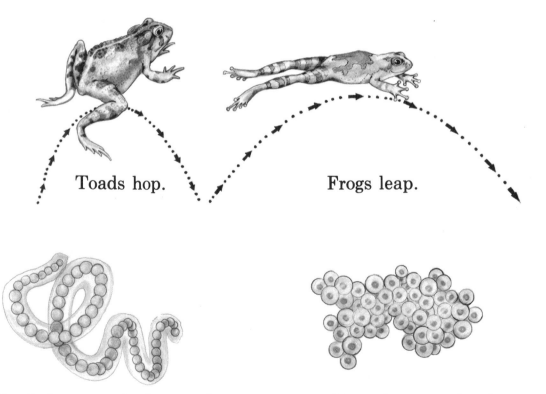

Toads hop.   Frogs leap.

Toads lay eggs in long strings.   Frogs lay eggs in clumps.

There is another way to find out
whether you are looking at a frog or a toad.

Only toads have large bumps (*paratoid glands*)
behind the eyes.
They are usually shaped like beans.

PARATOID GLAND

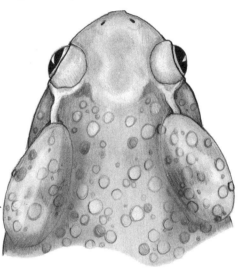

The Spadefoot toad is really a frog,
even though it is called a toad.
Can you tell why?

Look for bumps behind the eyes.
There aren't any.

It is easy to tell frogs from toads
but it is hard to tell toads apart.

You have to look closely at this bumpy creature.

AMERICAN TOAD

RED-SPOTTED TOAD

Which toad has round instead of bean-shaped bumps *behind* its eyes?

Which toad has a bump *between* its eyes?

Which toad has one or two warts in each dark spot?

Which toad has many more warts in each dark spot?

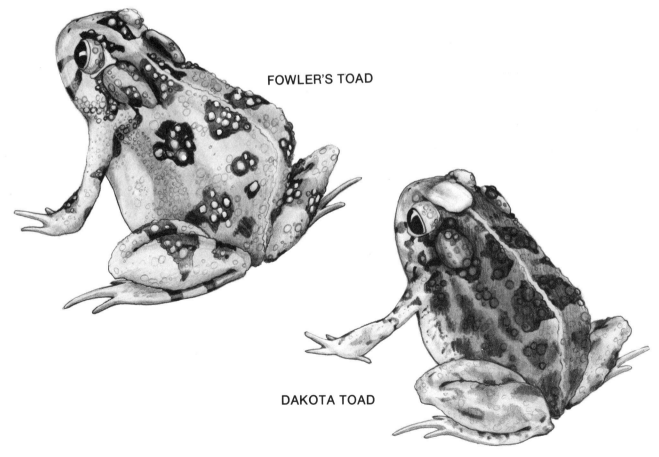

FOWLER'S TOAD

DAKOTA TOAD

The Great Plains toad has light rings
around its large, dark spots.

The Gulf Coast toad has a broad, dark stripe
on each side.

Which is which?

Salamanders are easy to tell from frogs and toads because they have tails.

Many people mix up salamanders with lizards.

But lizards have scales and claws.
Salamanders do not.

Salamanders come in different sizes.

A salamander can be over 20 inches long.

HELLBENDER

Or a salamander can be only 2 inches long.

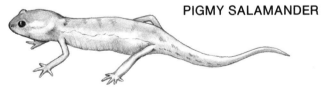

PIGMY SALAMANDER

Most salamanders are under 5 inches long.

Salamanders come in different shapes.

This one is short and chunky.

MARBLED SALAMANDER

This one is long and thin.

WORM SALAMANDER

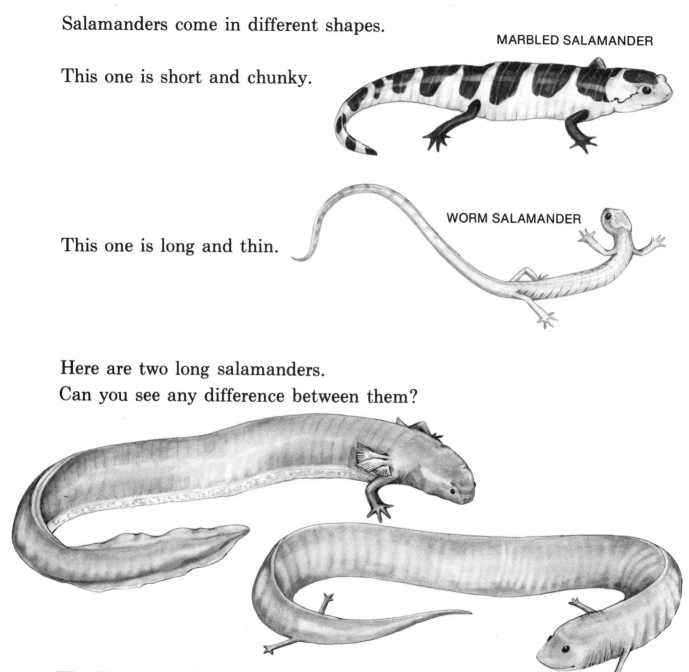

Here are two long salamanders.
Can you see any difference between them?

The Siren has gills. The Congo eel has none.
Do you see another difference? Hint: count the legs.

Salamanders have different skin patterns.

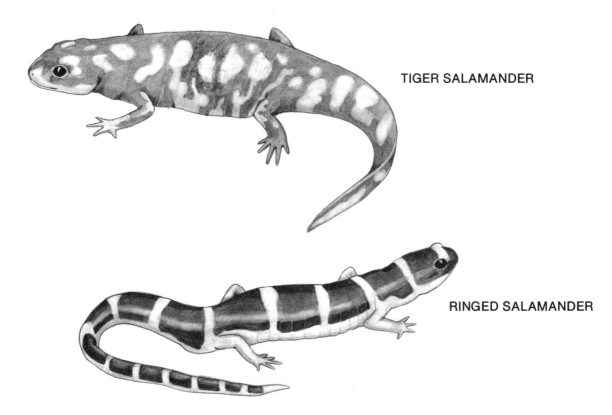

TIGER SALAMANDER

RINGED SALAMANDER

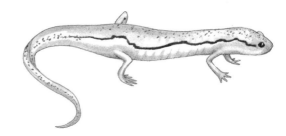

BLUE RIDGE SALAMANDER

Find the salamander with large, round spots.
Find the salamander with blotches.
Find the salamander with stripes across its body.
Find the salamander with stripes along each side.
Find the salamander with rows of small dots down its back.

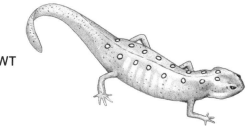

RED-SPOTTED NEWT

SPOTTED SALAMANDER

When you look at an amphibian
you have to notice many things.

Look at the shape.

Look at the pattern of its skin.

Look at the size.

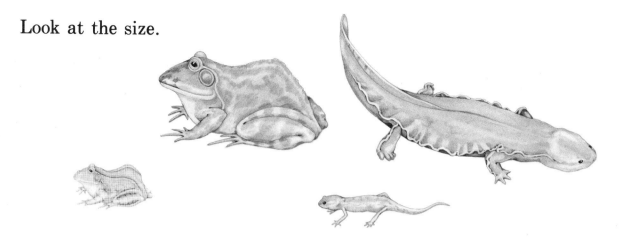

Look for bumps on toads.

Look at the head, body, feet and eyes.

31

Where to find the frogs, toads and salamanders
mentioned in this book:

Ponds and lakes—Bullfrog, Green Tree frog,
Green frog, Leopard frog, Tiger salamander, Ringed
salamander and Spotted salamander

Brooks, streams and rivers, and the meadows around them—
Pickerel frog, Dakota toad, Hellbender salamander

Marshes, muddy lakes and swamps—Siren, Congo eel,
Swamp Tree frog, Narrow-mouth frog

Open grasslands—Great Plains toad, Red-Spotted toad

Loose, sandy soil—Spadefoot toad, Sheep frog,
Marbled salamander

Woodlands—Spring peeper, Woodfrog, Gray Tree frog,
Pigmy salamander, Blue Ridge salamander

Caves—Barking frog

City back yards to mountains and forests—American toad,
Fowler's toad

5893 $5.39

597
SEL   Selsam, Millicent E.

A first look at
frogs, toads and
salamanders

597
SEL   5893   Selsam, Millicent E. $5.39

A first look at
frogs, toads and
salamanders

DATE	BORROWER'S NAME	
DE 18 '78	Mark Pitts	5 B
MR 18 '82	A Bradfield	
MR 15 '84	C. Bumphus	8-B
MR 16 '88	J F	4

DE 18 '7	
MR 18 '8	
MR 15 '	
MR 16	
MY 2	
MAR 2	
AP 30	